Star Portals & Celestial Journeys

Gnostic and Biblical Texts of Heavenly Ascents

From the Ladder of Jacob to Nag Hammadi's Cosmic Realms—Interstellar Travel in Sacred Writings

A Modern Translation
Adapted for the Contemporary Reader

Various Ancient Writers

Translated by Tim Zengerink

© **Copyright 2025**
All rights reserved.

It is not legal to reproduce, duplicate, or transmit any part of this document in either electronic means or in printed format. Recording of this publication is strictly prohibited and any storage of this document is not allowed unless with written permission from the publisher except for the use of brief quotations in a book review.

This book contains works of fiction. Any resemblance to persons living or dead, or places, events, or locations is purely coincidental.

Table of Contents

Preface - Message to the Reader 1

Introduction 5

The Ladder of Jacob 11

Ascension of Isaiah 19

    Chapter 1 19
    Chapter 2 20
    Chapter 3 23
    Chapter 4 26
    Chapter 5 28
    Chapter 6 30
    Chapter 7 32
    Chapter 8 35
    Chapter 9 37
    Chapter 10 41
    Chapter 11 44

Thank You for Reading 48

Preface - Message to the Reader

What If You Could Help Rebuild the Greatest Library in Human History?

Thousands of years ago, the Library of Alexandria stood as the crown jewel of human achievement — a sanctuary where the collected wisdom of every known civilization was gathered, preserved, and shared freely.

And then, it was lost.

Through fire, conquest, and the slow erosion of time, humanity lost not just books — but ideas, dreams, discoveries, and stories that could have changed the world forever.

Today, the Library of Alexandria lives again — and you are invited to be a part of its restoration.

Our mission is simple yet profound:

To rebuild the greatest library the world has ever known, and to translate all timeless works into every language and dialect, so that no seeker of knowledge is ever left behind again.

By joining our movement to rebuild the modern Library

of Alexandria, you become part of an unprecedented mission:

- **Unlimited Access to the Greatest Audiobooks & eBooks Ever Written:**

  Instantly explore thousands of legendary works—Plato, Shakespeare, Jane Austen, Leo Tolstoy, and countless more. All instantly available to read or listen, placing a complete literary universe at your fingertips.

- **Beautiful Paperback & Deluxe Editions at Printing Cost**

  Own any title as an elegant paperback, deluxe hardcover, or stunning collectible boxset—offered to you at true printing cost, delivered straight to your door. Build your personal Library of Alexandria, crafted for beauty, built for durability, and worthy of proud display.

- **Fresh Translations for Modern Readers—in Every Language & Dialect**

  Enjoy timeless masterpieces reimagined in clear, contemporary language—no more outdated phrases or obscure references. Alongside the original versions, we're tirelessly translating these classics into every language and dialect imaginable, ensuring accessibility and understanding across cultures and generations.

- **Join a Global Renaissance of Literature & Knowledge**

  You directly support expanding our library, publishing deluxe editions at true cost, translating works into all global languages, and bringing humanity's greatest stories to people everywhere. By joining today, you're not just preserving a legacy of masterpieces; you set in motion a powerful wave of literary accessibility.

Become a Torchbearer of Knowledge.

Join us for free now at **LibraryofAlexandria.com**

Together, we will ensure that the light of human wisdom never fades again.

With gratitude and a shared love of knowledge,
The Modern Library of Alexandria Team

Visit:

www.libraryofalexandria.com

Or scan the code below:

Introduction

Gateways Between Worlds: The Visionary Roots of Ascent Literature

Throughout history, certain sacred texts have dared to explore what lies beyond the veil of visible reality. While much of religious scripture anchors itself to terrestrial concerns—laws, ethics, the history of peoples—there exists another strand within the broader fabric of Jewish and Christian mystical literature: one that focuses on the upward journey of the soul, the structure of the heavens, and the soul's potential passage through divine realms. These works, rich in visionary experience and cosmological speculation, offer tantalizing glimpses into ancient spiritual worldviews that blur the line between mysticism, astronomy, metaphysics, and what some today might interpret as proto-science fiction.

This collection, Star Portals & Celestial Journeys, gathers some of the most powerful and enigmatic examples of that visionary tradition. Texts like the Ladder of Jacob, the Ascension of Isaiah, Zostrianos, the Paraphrase of Shem, and other celestial narratives transport readers across the heavens, describing majestic gateways, spiritual battles, and realms populated by luminous beings. For the faithful, these texts represent an invitation to mystical communion. For the curious, they raise profound questions: Are these symbolic visions of divine ascent? Astral meditations? Allegories of spiritual growth? Or could they be echoing something even more extraordinary—encounters with actual interdimensional phenomena?

Whether interpreted metaphorically or literally, these works hold enduring fascination. They propose that reality is far grander than the material world suggests—and that access to transcendent realms is possible, if not inevitable. These texts situate humanity not merely as inhabitants of Earth, but as beings poised on the edge of a multidimensional cosmos.

Ancient Ascent Narratives and Their Celestial Maps

Among the oldest and most striking features of these writings is their detailed portrayal of the universe as a structured, tiered cosmos. These were not vague notions of "heaven above," but intricate metaphysical cartographies, carefully outlined and populated with angels, archons, and heavenly guardians. The Ladder of Jacob, for instance, presents a dream in which Jacob beholds ascending and descending angelic beings—visions that echo the famed "stairway to heaven" in

Genesis, but with added metaphysical depth. This text situates Jacob not only as a patriarch but also as a proto-mystic, witnessing a structure that links the terrestrial and celestial realms.

The Ascension of Isaiah, one of the most detailed early Christian ascent texts, maps out seven heavens, each one containing its own hierarchies, dangers, and thresholds. It shares thematic and structural similarities with the Testament of Levi, 3 Enoch, and the mystical merkabah literature of the Hebrew tradition. The soul's journey through each level is fraught with spiritual challenges. In later Gnostic writings such as Zostrianos, the path is even more complex: layered with aeons, emanations, and zones of light and ignorance.

Modern readers may see in these descriptions early expressions of multi-dimensional theory or proto-concepts of wormholes and interstellar gateways. While that interpretation is necessarily speculative, it is worth noting that these ancient authors took great care in describing spiritual ascents with a precision that seems oddly congruent with what today's science fiction and astrophysics call higher-dimensional travel.

Indeed, the texts found among the Nag Hammadi codices—including Zostrianos, Marsanes, and the Paraphrase of Shem—are sophisticated philosophical treatises as much as they are mystical gospels. Their fusion of Neoplatonic metaphysics, Hebrew angelology, and Egyptian cosmic symbolism suggests an era when spiritual exploration was also a rigorous intellectual pursuit. Their vision of reality encompasses cosmic forces, vast realms of spirit and matter, and a Divine Mind beyond all knowing—reminding us that the ancients may have conceived of spiritual ascent in ways not dissimilar to modern notions of universal consciousness or the transcendence of physical form.

Modern Resonance:
Ancient Mysticism Meets Cosmic Curiosity

One of the most fascinating aspects of this collection is how naturally these texts intersect with contemporary curiosity about space, time, and interdimensional life. The notion of "portals" in these writings—passages between realms—has found strange parallels in modern physics and science fiction. Concepts such as wormholes, extra dimensions, and alternate realities are increasingly part of mainstream thought. For some readers, the journey of the soul through heavenly gates may echo quantum theories about multiverse traversal or dimensional ascension.

Yet the value of these texts does not lie solely in their speculative potential. They also offer timeless wisdom about the human soul's yearning for transcendence, the struggle against spiritual ignorance, and the hope of divine reunion. In mystical terms, the heavens through which the soul travels may not be spatial locations but states of being—refinements of consciousness that the seeker ascends through prayer, meditation, or initiation.

This dual capacity—to speak both mythologically and metaphysically—is what gives these texts their power. They can be read as maps of the cosmos or blueprints of the inner world. Their images are potent: chariots of fire, thrones of glory, veils of light, and hidden gates guarded by sentient beings. These are not just stories of otherworldly travel but meditations on transformation. The traveler—whether Jacob, Isaiah, Zostrianos, or the reader—must be purified, stripped of illusion, and prepared to encounter the Divine Mind.

Thus, these writings represent a sacred genre: the literature of ascent, where the material world is not denied but transcended. They urge us not to flee the world, but to see through it—to recognize it as a layer within a greater cosmic tapestry.

As you read Star Portals & Celestial Journeys, consider the texts as spiritual telescopes, aimed not only outward toward the heavens but inward toward the soul. Whether they are symbolic allegories or actual visionary records, they share a bold premise: that reality is multidimensional, and that the soul was meant to journey beyond. They invite us into mystery, not with fear, but with awe.

This book is not only a collection of ancient texts. It is a call to awaken—to seek the truth behind the stars, the gates beyond the firmament, and the divine presence that calls each of us home.

The Ladder of Jacob

Jacob went to stay with his uncle Laban. As the sun set, he found a spot to sleep, using a stone as a pillow. That night, he had a dream.

In the dream, he saw a ladder standing on the ground that reached all the way to the sky. At the top of the ladder was a glowing face that looked like it was made of fire and shaped like a person. The ladder had twelve steps, and on each step were two human faces—one on the left and one on the right—making twenty-four faces in total. But one face in the middle was bigger and brighter than the rest. It stretched from the shoulders to the arms and looked more powerful and frightening than any of the others.

As Jacob watched, he saw angels going up and down the ladder. Then he looked up and saw the Lord standing above it. The Lord called out, "Jacob, Jacob!" Jacob replied, "I'm here, Lord." Then the Lord said, "The land you are lying on will belong to you and your children after you. I will make your family as countless as the stars in the sky and the sand by the sea. Because of your family, all the people on earth will be blessed—even in the far future. The blessing I give you will continue for many generations. People from the east and west will be part of your family."

Hearing this made Jacob shake with fear. He woke up, still feeling the presence of God's voice. He said, "This place is holy! It must be God's house and the gate of heaven." Then he took the stone he had used as a pillow, stood it upright as a marker, poured oil over it, and named the place "God's House."

Then he prayed, "Lord God, Creator of Adam, God of Abraham and Isaac, my father, and of all who live faithfully—you sit above the cherubim and rule from a throne that glows with fire and is covered in eyes. You carry the cherubim with four faces, you support the seraphim with many eyes, and you hold the world in your arms while no one holds you. You created the heavens to honor your name and spread the skies above the clouds. You guide the sun and hide it at night so people don't think it's a god. You made a path for the moon and stars. The moon grows and shrinks, and the stars move just as you told them to—so they won't be mistaken for gods either.

The six-winged seraphim stand in front of you. They cover their faces and feet with their wings and fly with the others, singing, 'Holy, Holy, Holy!' You are the Most High with many names. You are fire and lightning, full of glory. You are Jao, Jaoel, Sabakdos, Chabod,

Sabaoth, Omlelech, Elaber, and more. You are the eternal King—strong, mighty, patient, and full of greatness. You fill the heavens, the earth, the sea, and even the deepest places with your glory. Please hear my prayer. Listen to my praise. Show me what my dream means. You are strong, holy, and full of glory. You are my God and the God of my fathers."

While Jacob was still praying, a voice appeared in front of him. It said, "Sarekl, leader of those who serve with joy and keeper of visions—go help Jacob understand the dream he saw. Show him everything he saw, but first, bless him."

Then the archangel Sarekl came to Jacob. His face looked powerful, but Jacob wasn't scared. The face he had seen in the dream was even more intense, so he wasn't afraid of the angel. The angel asked, "What is your name?" Jacob answered, "Jacob." Then the angel said, "From now on, you will no longer be called Jacob. Your name will be like mine—Israel."

Later, as Jacob traveled from Fandana in Syria to meet his brother Esau, Esau came to him, blessed him, and also called him Israel. Esau didn't tell Jacob his own name at first. Only when Jacob begged him did Esau finally share it and explained that it was connected to something Jacob had done. (This part seems to mix up

Jacob's dream with the story of when he wrestled the angel, and the details are a little unclear.)

The angel explained, "The ladder you saw with twelve steps and faces on each side stands for the world you're living in now. Each step is a period of time, and the twenty-four faces are the kings who rule over non-believing nations. During their rule, your descendants will suffer because of their sins. The place you saw will be destroyed four times. Still, the temple will be rebuilt to honor the God of your ancestors. But because of their sin, it will be ruined again, and will stay that way until the fourth generation has passed. That's why you saw four visions."

Then Jacob saw someone trip while climbing the ladder. He saw angels moving up and down and more faces along the steps. God would raise a leader from Esau's family. All the rulers of the nations who had hurt Jacob's people would give in to this leader. He would treat them harshly and rule over them with force. They wouldn't be able to stop him. Eventually, he would demand that people worship false gods and the dead. Many from Jacob's family would be handed over— some to this harsh ruler, and others to God's judgment.

Jacob, I want you to understand this: your descendants will one day live in a foreign land, where they will be treated badly and forced into slavery. They will be beaten and hurt daily. But the people who treat them this way will be judged by God. When the time comes and a king rises up and fights back, God will bring judgment to that place. Then your people, the children of Israel, will be freed from the power of those who ruled over them with cruelty. They will no longer be shamed or mocked by their enemies.

This king will be the one who brings justice and punishment to those who attacked Israel. At the end of this time, those who suffer will cry out, and the Lord will listen. He will feel compassion, and even the strongest will show pity because angels and archangels will pray for the rescue of your people. Women among your people will have many children again, and the Lord will fight for them.

The land of their enemies will suffer. Their food supplies will be empty, there will be no wine or fruit. The ground will be full of crawling creatures and other harmful things. There will be earthquakes and destruction. When God passes judgment on that land, He will lead your people out of slavery. They will be saved from the insults of their enemies.

The king who stood against them will face revenge. Even though he stood proudly, thinking he was powerful, the people cried out—and the Lord heard them. He poured out His anger on Leviathan, the sea monster, and struck down the wicked ruler named Thalkon with a sword because that ruler had lifted himself up against the true God.

Then, Jacob, your righteousness and that of your ancestors—and of your children who follow your path—will be seen. Your descendants will blow the trumpet, and the entire kingdom of Edom will be destroyed, along with the kings and people of Moab.

The dream you had about the angels going up and down the ladder points to something in the future. In the last days, someone will come from God. He will want to connect what is above with what is below. Before He arrives, your sons and daughters will speak messages from God, and young people will have visions of Him.

Strange and powerful signs will happen when He is about to come: a tree that is cut down will bleed, babies just a few months old will speak clearly, and a child still in the womb will announce His coming. A young man will seem wise like an old man. Then, the One who has long been expected will arrive, but no one will know exactly how He comes. When He appears, the earth will celebrate because heaven's glory has come down to it. What was once far above will now be close.

From your family, a royal leader will grow. He will rise up and destroy the power of evil. He will be a savior—not just for your people, but for the tired and broken among the other nations too. He will be like a cloud offering shade from the heat, covering the whole world. Without Him, the world would remain broken. He is the one who connects heaven and earth.

When He arrives, idols made of bronze, stone, and carved images will speak out for three days. They will tell wise people what is happening on the earth. And by a special star, those seeking Him will find their way. They will see Him walking on earth—this One whom even the angels cannot fully see in heaven.

Then, God Himself will appear in human form. He will be held in the arms of a regular person. He will bring new life to humankind. He will restore what was lost when Adam and Eve sinned. The lies and tricks of evil will be broken, and all false gods will fall down in shame because they were built on lies. They will no longer be able to rule or pretend to speak truth. Their power will be taken from them, and they will have no more honor.

The child who comes will take their strength away. He will fulfill the promise God made to Abraham. He will smooth out everything rough and bring peace. He will throw all wickedness into the deep sea and perform amazing miracles in both heaven and earth.

But this savior will be wounded—in the house of someone He loves. When He is hurt, the time of salvation and the end of all evil will be near. Those who hurt Him will be wounded in return, and their wound will never heal. But the One who was wounded will be worshiped by all creation. People from everywhere, including all nations, will place their hope in Him. Everyone who knows His name will never be ashamed. His strength and His life will never end.

Ascension of Isaiah

Chapter 1

In the 26th year of King Hezekiah's rule over Judah, he called for his only son, Manasseh. He brought Manasseh into the presence of the prophet Isaiah, son of Amoz, and also Josab, Isaiah's son. He wanted to pass down to them the truths about righteousness that he had witnessed himself—truths about God's final judgment, the punishments of hell, the ruler of this world, and the dark powers and spirits under him. He also shared what he had learned about the faith of the Beloved One. He had seen these things during the 15th year of his rule, when he had been seriously ill. Hezekiah handed over scrolls written by Samnas the scribe, and others written by Isaiah himself. These scrolls were meant for the prophets to keep. They included visions he had seen from inside the palace: the judgment of angels, the destruction of the world, the clothes of the holy people, their transformation, and the suffering and rising of the Beloved One. Isaiah had seen all this in the 20th year of Hezekiah's reign and had already passed the message to his son Josab. Now, as Hezekiah gave instructions, Josab was there beside him. Isaiah then spoke to King Hezekiah, not just in front of Manasseh, and said, "As surely as the Lord lives, and as surely as His Spirit speaks through me, all these commands and words will be ignored by your son Manasseh. Because of him, I will suffer and die painfully. "Sammael, also known as Malchira, will work for Manasseh and do everything he asks. Manasseh will choose to follow evil instead of truth.

"He will lead many people in Jerusalem and Judah away from the true faith. Evil will live in him, and through him, I will be cut in half." When Hezekiah heard this, he cried hard. He tore his clothes, put dirt on his head, and fell on his face in grief. Isaiah said to him, "Sammael's plan for Manasseh is already complete. Nothing you do can stop it." That same day, Hezekiah decided in his heart that he would kill his own son, Manasseh. But Isaiah told him, "The Beloved One has blocked your plan. What you want to do will not happen. This is part of the calling I have received, and I will share in the reward that belongs to the Beloved."

Chapter 2

After King Hezekiah died, his son Manasseh became king. But Manasseh didn't follow his father's teachings—he forgot all of them. Sammael, the evil spirit, stayed close to Manasseh and had a strong hold on him. Manasseh turned away from serving the God his father worshiped. Instead, he followed Satan, his demons, and their evil powers. He changed the house of worship that Hezekiah had used to honor God. He rejected wisdom and stopped serving the true God. Manasseh gave his heart to Beliar, the spirit of lawlessness and ruler of this world. Beliar's other name is Mantanbuchus. He was happy with what Manasseh was doing in Jerusalem, so he gave him power to lead more people into sin. The city became full of evil. There was more witchcraft, spells, fortune-telling, and evil signs. Sexual sin, cheating in marriage, and the abuse of good people increased. Manasseh, along with Belchira, Tobia the Canaanite, John from Anathoth, and Zadok the chief builder, all took part in these wrongdoings. Everything else Manasseh did is written in the Book of the Kings of Judah and Israel. When Isaiah, son of Amoz, saw how much evil was happening in Jerusalem and how people were worshipping Satan and doing whatever they wanted, he left the city. He moved to Bethlehem in Judah.

But even in Bethlehem, there was a lot of wickedness. So Isaiah left again and went to live alone in the mountains, in a desert place. There, other prophets joined him—Micaiah, the old prophet Ananias, Joel, Habakkuk, Isaiah's son Josab, and many others who believed that people could rise up to heaven. They all settled on the mountain together. They wore rough clothes made from animal hair and had nothing else. They lived like this and cried loudly because the people of Israel had turned away from God. They only ate wild plants they picked from the mountains. They cooked these plants and lived on them while staying with Isaiah. They stayed on the hills and mountains like this for two years. While they were still living in the desert, a man named Belchira showed up. He was from Samaria and belonged to the family of Zedekiah, son of Chenaan. Belchira was a false prophet who had lived in Bethlehem. His relative, Hezekiah son of Chanani, had taught the 400 prophets of Baal during the time of Ahab, king of Israel. He had once hit and insulted the true prophet Micaiah, son of Amada. Micaiah had been punished by Ahab and thrown into jail. He had also been with Zedekiah the prophet, and both of them were around during the time of Ahaziah, son of Ahab, who ruled in Samaria. Elijah the prophet, from Tebon in Gilead, warned Ahaziah and the people of Samaria. He said that Aḥaziah would die in bed and that Samaria would be taken over by Leba Nasr because they had killed God's prophets. When the false prophets who followed Ahaziah—and their teacher Jalerjas from Mount Joel—heard this… Jalerjas, who was Zedekiah's brother, convinced King Ahaziah of Aguaron to have Micaiah killed.

Chapter 3

Belchira found out where Isaiah and the prophets were staying. He lived near Bethlehem and supported King Manasseh. He was a false prophet in Jerusalem, and many people there agreed with him. He was from Samaria. When King Alagar Zagar of Assyria came and captured people, taking them to the mountains of the Medes and the rivers of Tazon, Belchira was still young. He had escaped and gone to Jerusalem during King Hezekiah's time. But even though he feared Hezekiah, he didn't live the right way. During Hezekiah's rule, Belchira was caught speaking against God in Jerusalem. The king's servants reported him, so he ran away to Bethlehem. There, people influenced him to turn against Isaiah and the other prophets. Belchira accused Isaiah and the others, saying, "Isaiah and his group are speaking against Jerusalem and all the cities of Judah. They say these places will be destroyed and that the people of Judah and Benjamin will be taken as prisoners. They even say terrible things will happen to you, my king—that you'll be dragged away in chains." He went on, "Their messages about Israel and Judah are false. Isaiah even said, 'I see more than Moses saw.' But

Moses said no one can see God and live. Yet Isaiah claims, 'I saw God and I'm still alive.'" Belchira added, "So, King, Isaiah is lying. He called Jerusalem 'Sodom' and said the leaders of Judah are like the people of Gomorrah." He brought all these accusations before Manasseh. Manasseh and his officials, even the princes of Judah and Benjamin, the king's servants, and counselors, all had Beliar—the evil spirit—working in their hearts. They were pleased with what Belchira said, so they arrested Isaiah. Beliar was especially angry at Isaiah because of his visions. Isaiah had revealed the truth about Sammael and had spoken of the coming of the Beloved from the seventh heaven—how He would come down, take human form, be mistreated, and suffer. Isaiah had prophesied that the Beloved would be tortured by the people of Israel, crucified before the Sabbath with criminals, and buried in a tomb. He also said the Beloved's twelve followers would be shaken by what happened. Guards would watch over His tomb, but an angel from the heavenly church would be sent in the final days. Isaiah said Gabriel, the angel of the Holy Spirit, and Michael, the leader of the angels, would come on the third day, open the tomb, and the Beloved would rise up on their shoulders. He would then send out His twelve disciples.

These disciples would go on to teach people from all nations and languages about His resurrection. Those who believed in His cross would be saved, and they would also believe in His return to the seventh heaven, where He came from. Many of His followers would speak through the Holy Spirit, and amazing miracles and signs would happen. But before He returned, Isaiah said, many of His followers would stop obeying the teachings of the Twelve Apostles. They would turn away from true faith, love, and purity. Arguments and divisions would be everywhere before His arrival. People would desire leadership roles even if they had no wisdom. Many corrupt elders and pastors would mistreat the people they were supposed to guide. They would be greedy and selfish. Holy clothing would be replaced by outfits worn for show, and people would only care about status and worldly honor. There would be lots of gossip, pride, and showing off. The Holy Spirit would leave many people. At that time, there would be very few real prophets. Only a few in different places would speak God's truth, because of the spirit of lies, lust, pride, and greed among people who claimed to serve the Beloved. Even the leaders would hate each other and fight. Jealousy would be everywhere. People would say whatever they felt like, and no one would respect the messages of earlier prophets—or even

Isaiah's visions. Instead, they would just follow their own ideas.

Chapter 4

Now listen, Hezekiah and my son Josab—these are the final days of the world. After everything is completed, Beliar, the powerful ruler of this world, will come down. He has ruled ever since the beginning. He'll come down from the skies looking like a man, a cruel king, one who even kills his own mother. That king will be Beliar himself. He will go after the group that the Twelve Apostles of the Beloved started. One of the twelve will fall into his hands.

This evil ruler, appearing as a king, will be followed by all the forces of this world. They will do everything he says. He'll perform strange signs. For example, the sun will rise during the night and the moon will shine at noon. He will do whatever he wants and speak like the Beloved. He'll say, "I am God. There's never been anyone before me." And most people in the world will believe him. They'll worship him and say, "This is the only God." Even many people who once followed the true Beloved will be tricked into following him instead. He'll perform powerful miracles in every city and region. He'll set up statues of himself everywhere. His rule will last for three years, seven months, and twenty-seven days. During that time, only a few true believers will remain—those who still hope in the One who was crucified, Jesus Christ the Lord. These believers, the ones who saw Him rise to heaven, will survive by running from one desert to another, waiting for the Beloved to return. Then, after 1,332 days, the Lord will come back with His angels and the armies of the holy ones from the seventh heaven. He will come in the glory of that highest heaven and throw Beliar and all his followers into hell. He will give peace and rest to those who stayed faithful, who still lived in their bodies on earth. Even the sun will be ashamed by His glory. He will reward those who stood strong in their faith and rejected Beliar and his evil rulers. The saints, wearing their heavenly clothes that have been kept for them in the seventh heaven, will return with the Lord. Their spirits will be dressed in glory, and they'll come down to earth. The Lord will strengthen the believers who are still alive, dressing them in the same holy clothes. He will care for those who stayed faithful. Then those holy ones will rise back up, leaving their earthly bodies behind.

The voice of the Beloved will shout in anger at everything—heaven, earth, the mountains, the cities, the forests, the sun and moon—everything that Beliar used to show off his power. All of it will be judged. Fire will come out from the Beloved. It will burn up all the wicked, and it will be as if they never existed. The rest of this vision is written in the scroll about Babylon. More about the Lord is written in three stories I told in the book where I shared these prophecies out loud. The part about the Beloved going down into the place of the dead is in the section where the Lord says, "Look, my Son understands." All of these things are also written in the Psalms, in the poems of David son of Jesse, in the Proverbs of Solomon his son, and in the songs of Korah, Ethan the Israelite, and Asaph. Other Psalms without names were written by the angel of the Spirit. They are also found in the words of my father Amos, and in the books of Hosea, Micah, Joel, Nahum, Jonah, Obadiah, Habakkuk, Haggai, Malachi, Joseph the Just, and Daniel.

Chapter 5

Because of these visions, Beliar became furious with Isaiah. He entered the heart of Manasseh, who then had Isaiah cut in half with a wooden saw. While Isaiah was being sawed, Belchira stood by accusing him. The false prophets were all there, laughing and celebrating Isaiah's suffering. Belchira, along with Mechembechus, mocked Isaiah and said to him, "Say this: 'I lied about everything I said. Manasseh's ways are right and good.

Also say, 'The ways of Belchira and his friends are good too.'" He told Isaiah to say this just as the sawing began. But Isaiah was caught up in a vision from the Lord. His eyes were open, but he didn't see the people around him. Belchira tried again, saying, "Say what I tell you, and I'll change their minds. I'll make Manasseh, the leaders of Judah, the people, and all of Jerusalem respect you."

Isaiah replied, "As long as I can speak, I say this: You and your powers and your whole house are cursed. You can only harm the skin of my body—you can't touch anything more." Then they grabbed Isaiah and sawed him in half with a wooden saw. Manasseh, Belchira, the false prophets, the leaders, and all the people stood there watching. Before Isaiah was cut in two, he told the prophets who were with him, "Go to the land of Tyre and Sidon. This suffering is mine alone. God has chosen this cup for me." As Isaiah was being sawed, he didn't scream or cry. He quietly spoke through the Holy Spirit until his body was split in two. Beliar used Belchira and Manasseh to do this to Isaiah because Sammael had been angry with him since the days of King Hezekiah. Isaiah had seen visions of the Beloved and the defeat of Sammael through the Lord. So they carried out Satan's plan.

Chapter 6

This is the vision that Isaiah, son of Amoz, saw: In the twentieth year of King Hezekiah's reign in Judah, Isaiah and his son Josab traveled from Galgala to Jerusalem to meet with Hezekiah. When they arrived, Isaiah sat down on the king's couch. Someone brought him a seat, but he refused to sit on it. Isaiah began to speak words of faith and truth to King Hezekiah. All the royal officials, advisors, and forty prophets—including the sons of prophets—were there. They had come from villages, mountains, and the plains because they heard Isaiah was coming to speak with the king. They came to show him respect, to hear his words, and to receive his blessing. They hoped he would lay hands on them so they could also speak prophetic words, and so he could listen to their messages from God. Everyone gathered around Isaiah.

While Isaiah was still speaking words of faith to Hezekiah, they all heard a door open and the voice of the Holy Spirit. The king called all the prophets and everyone nearby to come and listen. They came, including Micaiah, the old prophet Ananias, Joel, and Josab, who sat beside him on both sides. When everyone heard the voice of the Holy Spirit, they dropped to their knees and worshipped. They praised the God of truth, the Most High who rules from above, the Holy One who lives among His holy servants. They honored God for giving a doorway between this world and another, and for choosing a human being to receive such a gift. As Isaiah spoke through the Holy Spirit, he suddenly stopped. His thoughts were taken from him, and he no longer saw the people in front of him. Even though his eyes were open, he didn't speak, and his thoughts were no longer with his body. But he was still breathing, because he was seeing a vision. The angel who came to show him the vision wasn't from this world or one of the normal angels here. This angel came from the seventh heaven. The people nearby didn't understand what was happening, but the circle of prophets realized that Isaiah had been lifted into a spiritual vision. The vision he saw wasn't from this world—it came from a place that ordinary humans can't see. After the vision ended, Isaiah told King Hezekiah, his son Josab, and the other prophets what he saw. But the rulers, eunuchs, and most of the people didn't hear the vision. Only Samna the scribe, Ijoaqem, and Asaph the recorder heard it. They were also righteous men, and God's Spirit was with them. The rest of the people didn't hear because Micaiah and Josab had led them away once Isaiah's spirit had been taken up. Isaiah looked as if he were dead because all worldly understanding had left him.

Chapter 7

This is the vision that Isaiah shared with Hezekiah, his son Josab, Micaiah, and the other prophets: Isaiah said, "While I was speaking the message you heard, I saw a shining angel. He was more glorious than any angel I had ever seen before. His greatness and brightness were beyond anything I can explain. He took my hand and lifted me up. I asked him, 'Who are you? What's your name? Where are you taking me?' I had been given strength to speak with him. He answered, 'Once I've taken you up through the different levels of heaven and shown you the vision I was sent to reveal, then you'll understand who I am. But you won't learn my name, because you'll be returning to your body. Still, you'll see where I'm taking you, because that's the reason I was sent.' I was happy because he spoke kindly to me. He asked, 'Are you glad because I spoke kindly? Just wait—you'll see someone even greater than I am speak to you gently and peacefully. You'll even see His Father, who is greater still. That's why I was sent from the seventh heaven—to show you all this.' Then he took me up to the sky, and I saw Sammael and his army. There was a huge battle happening. The angels of Satan were fighting each other, full of jealousy. It was just like what happens here on earth—what I saw in the sky is mirrored in this world. I asked the angel, 'Why are they fighting? What's all this jealousy about?' He said, 'It's been this way ever since the world was created. And it will continue until the One you're going to see comes and destroys him.' Then he lifted me above the sky to the first heaven. There, I saw a throne in the middle, with angels on the right and left. The angels on the right were brighter and more glorious than those on the left. They praised together with one voice. Then the ones on the left also gave praise, but their voices weren't as strong or beautiful. I asked the angel, 'Who are they praising?'

He replied, 'They're praising the One who sits in the seventh heaven—He who lives in the holy world—and they also praise His Beloved, the One who sent me to you.' Then the angel took me to the second heaven. It was as far above the first as the first is above the earth. There, just like before, were angels on both sides and a throne in the middle. The one sitting on that throne was even more glorious than all the others. The praise in the second heaven was greater than in the first. I fell down to worship, but the angel stopped me. He said, 'Do not worship any throne or angel in the first six heavens. That's not why I brought you here. I'm to guide you to the seventh heaven, where you'll learn everything.' He told me, 'Your throne, crown, and special clothes are already waiting for you in the highest heaven.' I was filled with joy, knowing that those who love the Most High and His Beloved will also rise up through the angel of the Holy Spirit. He brought me up to the third heaven. Like before, there were angels on the left and right, and a throne in the center. But here, no one spoke of the world below. I said to the angel, because I noticed my appearance changing as we rose higher, 'There's no mention of earthly things here.' He answered, 'Earthly things aren't mentioned here because they're weak. Nothing that's done is hidden here.' I wanted to understand how everything was known. The angel said, 'When I take you to the seventh heaven—the place I was sent from—you'll see that nothing is hidden from the thrones, the angels, or the ones who live there. The praise and glory in that place is even greater than what we've seen so far.' Then he took me to the fourth heaven. The distance from the third to the fourth was even greater than from earth to the sky. Again, I saw angels on both sides of a throne. They were praising the one who sat on it. The angels on the right shined more brightly than those on the left. And the one on the throne was more glorious than all the angels. His brightness surpassed even those who were already above the heavens below.

He then brought me to the fifth heaven. There, once again, I saw angels on both sides and a great throne in the middle. The glory of the one sitting on it was even greater than the one in the fourth heaven. The angels on the right shined brighter than those on the left, just like before. And the one on the throne was even more glorious than them all. The praise in the fifth heaven was louder and more powerful than in the fourth. And I praised the One whose name no one knows—the Only Son who lives in the heavens—who gave such beauty and power to each level of heaven, and who made the angels and the thrones even more glorious.

Chapter 8

Then he took me up into the air of the sixth heaven, and I saw more beauty and brightness than in all the five heavens below. I saw angels shining with amazing glory. Their singing and praise were holy and breathtaking. I asked the angel guiding me, "What am I seeing, my lord?" He answered, "I'm not your lord—I'm just your fellow servant." I asked again, "Why aren't there any angels on the left side?" He replied, "From this level upward, there are no more angels on the left. There's no throne in the middle either. Everything here is directed by the power of the seventh heaven, where the One who cannot be named lives, along with the Chosen One. No one here knows His name. None of the heavens above or below know it either. He alone speaks, and all the heavens and thrones answer Him. I was given the power and sent here to lift you up so you could see this glory, And so you could see the Lord who rules all the heavens and thrones. You'll watch as He changes form until He looks just like you.

Let me tell you something, Isaiah: no human who plans to return to their body has ever seen what you're seeing, or will see what you're going to see. This is a special gift. You were chosen by the Lord's will to come here." And I gave praise to my Lord, thanking Him that I was chosen to be brought here. Then the angel said, "Listen to something else, my fellow servant: once you leave your body, as God allows, and come here again, you'll receive the special clothing you see here—along with other garments kept for you—and You will be like the angels of the seventh heaven." He lifted me even higher into the sixth heaven again, and just as before, there were no angels on the left side, no throne in the middle. Everyone looked the same and sang praises equally. I was also given power to praise, and I joined the angel in singing with them. Our voices matched theirs. All together, they praised the Eternal Father, His Beloved—the Christ—and the Holy Spirit with one voice. Their singing wasn't like anything I'd heard in the lower heavens. Their voices and words were completely different, and there was a brilliant light everywhere. And being in the sixth heaven made all the light I had seen in the lower heavens seem like darkness in comparison. I was filled with joy and praised the One who gave such light to those who wait for His promises. I begged the angel who guided me not to send me back to the physical world. I tell you the truth, Hezekiah, Josab my son, and Micaiah—there is so much darkness down there. The angel guiding me knew what I was thinking and said, "If this light fills you with joy now, just wait until you see the light in the seventh heaven—where the Lord and His Beloved are, the One who sent me.

He will be called the 'Son' in your world, even though He hasn't appeared there yet. You'll see the special garments, thrones, and crowns waiting for the good people—those who put their hope in the Lord who will come down in your human form. The light there is greater and more amazing than you can imagine. But as for your wish to stay here, your time on earth isn't over yet." When I heard that, I felt sad. But he said, "Don't be upset."

Chapter 9

Then he took me up into the air of the seventh heaven. I heard a voice ask, "How far will someone in a human body be allowed to go?" I became very afraid and started shaking. But then I heard another voice saying,

"Isaiah is allowed to enter here, because this is where his robe is waiting." I turned to the angel with me and asked, "Who tried to stop me, and who allowed me to go up?" He replied, "The one who tried to stop you is in charge of the praises in the sixth heaven. The one who gave you permission is your Lord God, the Lord Christ. In the world, He will be called 'Jesus.' But you can't hear His real name until you leave your body." Then he brought me into the seventh heaven. I saw a beautiful, blinding light and too many angels to count. There, I saw holy Abel and all the good people who had lived righteously. I saw Enoch and the ones with him. They had taken off their earthly bodies and now wore heavenly robes. They looked like angels, standing in brilliant light. But none of them were sitting on their thrones, and none wore crowns. I asked the angel, "Why do they have robes but not thrones or crowns?" He answered, "They will receive their thrones and crowns after the Beloved comes down to Earth in a body like yours. In the last days, the Lord—who will be called Christ—will take on human form.

They already know who will get which throne and crown when He arrives. But when people see Him, they will think He's just a man made of flesh. The ruler of that world will attack Him. They will nail Him to a tree and kill Him, not realizing who He really is. Even the heavens won't understand what's happening—His true identity will be hidden. After He defeats the angel of death, He will rise on the third day and stay on Earth for 545 more days. Then many good people will rise with Him. Their spirits won't receive their robes until Christ rises, and they rise with Him. After that, they'll finally get their robes, thrones, and crowns when He enters the seventh heaven." Then I asked the same thing I had asked in the third heaven: "Show me how the things happening on Earth are known here." While I was still talking, one angel stepped forward. He was even more glorious than the one guiding me from the world. He showed me a book—not like the books from Earth—and opened it. He handed it to me. I read it and saw everything the people of Israel had done, along with others I didn't recognize, including my son Josab. I said, "It's true—nothing that happens on Earth is hidden in the seventh heaven." I also saw many robes, thrones, and crowns stored there. I asked the angel, "Who are these for?" He said, "These are for the people on Earth who will believe in the One I told you about. They will follow His words and believe in Him—and in His cross. These rewards are prepared for them." Then I saw someone standing whose glory was greater than anyone else's. His beauty was beyond anything I had seen. After I saw Him, all the good people and angels I had seen before came to Him. Adam, Abel, Seth, and the rest of the righteous came first. They worshipped Him, and all sang His praise together. I joined in too, and my voice sounded like theirs.

Then the angels came close, worshipped Him, and sang praises. I was changed again and became like one of the angels. The angel guiding me said, "Worship Him." And I did, giving Him praise. Then he told me, "This is the Lord of all the praises you've seen." While he was talking, I saw another glorious figure who looked just like Him. The righteous people went to worship this one too. I praised along with them, but this time, my appearance didn't change like theirs. Then the angels came and worshipped Him too. I saw the Lord and the second figure standing side by side. The second one stood to the left of the Lord. I asked, "Who is that?" The angel said, "Worship Him too. He is the angel of the Holy Spirit. He speaks through you and through the rest of the good people." I saw a powerful light with my spirit's eyes, but I couldn't look directly at it. Not even the angel with me, or any of the other angels worshipping the Lord, could look at it. But the righteous ones were able to see Him clearly. Then the Lord and the angel of the Spirit came close to me. The Lord said, "Look! You've been allowed to see God. And because of you, power has been given to the angel guiding you." I saw the Lord and the Spirit's angel worshipping together. They praised God with one voice. After that, all the good people came closer and worshipped too. The angels followed, singing praises with them.

Chapter 10

Then I heard the sounds and songs of praise that I had heard in each of the six heavens. They rose up and were heard even here.

All those praises were being sent to the Glorious One—whose beauty and power were too great for me to look at. But I could still hear and watch the praise being given to Him. The Lord and the angel of the Spirit were watching and listening too. All the worship from the six heavens wasn't just heard—it was also seen. Then the angel guiding me said, "This is the Most High, above all others. He lives in the holy world and rests among the holy ones. The righteous will call Him 'Father of the Lord' through the Holy Spirit." Next, I heard the voice of the Most High, the Father of my Lord, speaking to Christ—who will be called Jesus: "Go down through all the heavens. Go all the way to the sky above the world, and even to the angel in the place of the dead. But do not go to Haguel. Make yourself look like the beings in the five heavens. Be careful to look like the angels of the sky and even those in the underworld. None of the angels in that world will know you came from the seven heavens with Me. They won't realize who you are until I call out loudly to all the heavens, their angels, and their lights—even those in the sixth heaven. Then you will judge and destroy the rulers, angels, and gods of that world—because they denied Me and said, 'We are the only ones. There is no one else.' After that, you will rise from the place of death and return to your place. You won't need to change your shape in each heaven. Instead, you'll go back in glory and sit at My right hand. Then the rulers and powers of that world will bow down to You." These were the commands I heard the Great Glory give to my Lord. Then I watched my Lord leave the seventh heaven and enter the sixth. The angel guiding me said, "Watch closely, Isaiah. Now you'll see the Lord change His form and begin His descent."

I looked. As He entered the sixth heaven, the angels there saw Him and praised Him. He hadn't changed His shape yet, so they could tell He was different. I joined in their praise. But when He went down into the fifth heaven, He changed to look just like the angels there. Because of that, they didn't recognize Him, so they didn't praise or worship Him. Then He went down to the fourth heaven and again changed His shape to match the angels there. They also didn't praise Him, because He looked just like one of them. He kept going, down into the third heaven. He looked like the angels there too. The angels at the gate asked for a password, and the Lord gave it to them so they wouldn't recognize Him. Since He looked like them, they didn't praise Him either. Next, He descended to the second heaven. Once again, He gave the password at the gate and changed His form to match theirs. They didn't praise Him, because they thought He was just one of them. Then He went to the first heaven. There, too, He gave the password to the gatekeepers and made Himself look like the angels on the left side of the throne. They didn't praise Him either, for the same reason—He looked just like them. As for me, no one asked me anything because I was with the angel guiding me. Then He descended into the sky, where the ruler of this world lives. He gave the password to those on the left. He took on their form too, so they didn't praise Him. Instead, they were jealous of each other and constantly fighting. That place was full of evil and conflict over small things. Finally, I saw Him go down into the air, where He made Himself look just like the angels there. He didn't give a password this time, because those angels were too busy hurting and stealing from each other.

Chapter 11

After this, I saw more, and the angel who was guiding me said, "Isaiah, son of Amoz, understand this—God sent me to show you these things." Then I saw a woman named Mary. She came from the family of the prophet David. She was a virgin and engaged to a man named Joseph, who was a carpenter. He also came from David's family in Bethlehem, in the land of Judah.

When Joseph had taken his place in life, Mary, though still a virgin, was found to be pregnant. Joseph thought about ending the engagement quietly. But the angel of the Spirit appeared in this world. After that, Joseph didn't leave Mary. He stayed with her and didn't tell anyone what happened. Joseph didn't sleep with Mary. He kept her holy and untouched, even though she was expecting a child. For two months, they lived separately. Then one day, when they were alone together, Mary suddenly looked down and saw a baby. She was shocked. Right after that, her body appeared just as it had been before she became pregnant. Joseph asked, "What surprised you?" Then his eyes were opened, and he saw the baby too. He praised God because God had given him this gift. A voice told them, "Don't tell anyone about this vision." Still, word about the baby spread around Bethlehem. Some people said, "Mary had a baby before she was even properly married for two months." Others said, "No, she didn't. No midwife was with her. We never even heard her cry out in pain." They were confused about Him, and even though they knew about Him, they didn't understand where He had come from. Then they took the child and went to Nazareth in Galilee. I, Isaiah, saw this, Hezekiah and Josab, my son. I tell this also to the other prophets standing with us: all of this was hidden from the heavens, the rulers, and the false gods of this world. I saw Him in Nazareth, being nursed like any normal baby, so He wouldn't be noticed. When He grew up, He did many miracles and wonders throughout Israel and Jerusalem. But the enemy grew jealous and stirred up the people of Israel against Him. They didn't know who He really was. They handed Him over to the king, and He was crucified. He went down to the angel in the place of the dead.

In Jerusalem, I saw Him being nailed to a tree. Then, on the third day, He rose again and remained for a time. The angel guiding me said, "Understand this, Isaiah." And I saw Him send out the Twelve Apostles and rise into the sky. I saw Him in the sky above the world. He had not changed His appearance to look like the angels there. The angels of the sky and even the Satans saw Him and worshiped Him. There was great sadness among them. They asked, "How did our Lord come down among us, and we didn't notice the glory He had—the glory we now see that He brought from the sixth heaven?" Then He rose into the second heaven. He didn't change His form there either. All the angels to His right and left and those near the throne worshiped and praised Him. They asked, "How did our Lord pass through without us noticing?" He did the same in the third, fourth, and fifth heavens. They all said the same thing. His appearance didn't change, and the same glory stayed with Him the whole time. I saw Him rise to the sixth heaven. They worshiped Him there too, and the praise became even louder in each heaven. Then I saw Him enter the seventh heaven. All the righteous people and angels praised Him there. I saw Him sit down at the right hand of the Great Glory—whose light I told you I couldn't even look at. On the left side, I saw the angel of the Holy Spirit sitting. That angel said to me, "Isaiah, son of Amoz, you've seen enough. You've been shown things no other human being has ever seen. Now you will return to your earthly body until your time is complete. After that, you will return here." Isaiah shared everything he saw with those around him, and they praised God. He spoke to King Hezekiah and said, "I've told you everything." He spoke of the end of the world.

He said that everything he saw in the vision would come true in the last days. Isaiah made Hezekiah promise not to tell the people of Israel, or let anyone write down what he had said. "But one day," he said, "you will read about these things. Stay alert in the Holy Spirit so that you may receive your heavenly clothes, thrones, and crowns that are waiting in the seventh heaven." Because of these visions and prophecies, Satan (called Sammael) had Isaiah, son of Amoz, the prophet, cut in half by King Manasseh. Hezekiah gave these things to Manasseh in the twenty-sixth year of his reign. But Manasseh didn't remember or take them seriously. Instead, he served Satan and was destroyed. This is the end of the vision of the prophet Isaiah and his journey to heaven.

Thank You for Reading

Dear Reader,

We hope this timeless classic has sparked your imagination and enriched your literary journey. Now that you've turned the final page, we want to share a vision for the future of reading—one where every classic you've ever wanted to explore is at your fingertips, in a format that best suits your life.

We'd like to invite you to gain immediate, unlimited digital & audiobook access to hundreds of the most treasured literary classics ever written—along with the option to secure deluxe paperback, hardcover & box set editions at printing cost. Together, we can spark a new global literary renaissance alongside our small, independent publishing house called "The Library of Alexandria."

Thousands of years ago, the Library of Alexandria stood as a beacon of knowledge—until it was lost to history. We aim to reignite that spirit of preservation and discovery right now, in the modern age—only this time, it's accessible to all, in every language and every format.

Picture a world where every timeless classic, novel, poem, or philosophical treatise is not only available to read but also updated for today's readers—modernized, translated into any language or dialect, and ready to enjoy in any format you choose, whether that is in an eBook, audiobook, paperback, or deluxe hardcover & box set version a printing cost.

By joining our movement to rebuild the modern Library of Alexandria, you become part of an unprecedented mission to offer:

- **Unlimited Audiobook & eBook Access to the Greatest Classics of All Time**

    Instantly explore thousands of legendary works, from Plato and Shakespeare to Jane Austen and Leo Tolstoy. All are instantly ready to read or listen to, giving you a complete literary universe at your fingertips.

- **Paperback & Deluxe Editions at Printing Costs:**

    Purchase any title in a paperback, deluxe hardbound, or deluxe boxset edition at printing costs, shipped right to your doorstep. Curate your personal library of Alexandria with editions worthy of display—crafted to last, designed to captivate, and delivered straight to your door.

- **Modern translations for Contemporary Readers in all languages and dialects**

  Discover a vast selection of classics reimagined in clear, current language—no more struggling with outdated phrases or obscure references. Next to the original versions, we aim to offer translations in as many languages and dialects as possible.

  As we continue our translation efforts and add new languages, readers everywhere can connect with these works as if they were written today. By bridging linguistic divides, you're contributing to ensuring that these timeless stories become more meaningful, accessible, and inspiring for people across the globe.

- **Your Personal Library of Alexandria:**

  Over the months and years, you'll curate a unique physical archive of classics—each volume a testament to your taste, curiosity, and love of knowledge. It's not just about owning books—it's about curating a cultural legacy you'll cherish and pass down for generations to come.

- **Join a Global Literary Renaissance:**

  Your support fuels an ongoing mission: allowing us to reinvest in offering deluxe print editions (including special boxsets) at their true cost, broaden the range of available formats and translations, and extend the reach of these works to new audiences worldwide. By joining today, you're not just preserving a legacy of masterpieces; you set in motion a powerful wave of literary accessibility.

We are more than a publisher—we're a movement, and we can't do it alone. Your support lets us scale our mission, preserving and reimagining history's greatest works for tomorrow's readers.

Become a Torchbearer of knowledge.

Thank you for picking up this book and allowing us into your literary journey. As you turn the pages, know that you're part of something larger: a global effort to keep these stories alive, share their wisdom across borders and generations, and spark a true cultural revival for the modern era.

If this resonates with you—please consider taking the next step by visiting:

www.libraryofalexandria.com

With gratitude and a shared love of knowledge,

The Modern Library of Alexandria Team

Visit:

www.libraryofalexandria.com

Or scan the code below:

www.ingramcontent.com/pod-product-compliance
Lightning Source LLC
LaVergne TN
LVHW030631080426
835512LV00021B/3458